FROM A TREE TO A CHAIR

Written by Roseanne McDonald
Illustrated by Mary Barrows

SPRINGDALE PUBLIC LIBRARY

BELLE ISLE BOOKS
www.belleislebooks.com

Copyright © 2020 by Roseanne McDonald.

No part of this book may be reproduced in any form or by any electronic or mechanical means, or the facilitation thereof, including information storage and retrieval systems, without permission in writing from the publisher, except in the case of brief quotations published in articles and reviews. Any educational institution wishing to photocopy part or all of the work for classroom use, or individual researchers who would like to obtain permission to reprint the work for educational purposes, should contact the publisher.

ISBN: 978-1-947860-76-6

LCCN: 2019911804

Designed by Abigail Montiel

Project managed by Abigail Montiel and Haley Simpkiss

Printed in the United States of America

Published by

Belle Isle Books (an imprint of Brandylane Publishers, Inc.)

5 S. 1st Street
Richmond, Virginia 23219

belleislebooks.com | brandylanepublishers.com

Listen Intently
Cherish Every Moment

This book is dedicated to my son,
who loved nature and all things of wonder.

Everest James Cooper McDonald
May 22, 1979 - July 31, 2013

↖ forest in Oregon

Trees Growing in the Forest

Deep in the forest, it is usually very quiet. But this area is noisy with the loud *beep-beep-beeps* of a yarder, the *cling-clang-cling* from its cables, and the busy *rev-rev-REV* of a chain saw. Can you guess what is happening? Trees are being harvested!

Wood from trees is used for many things, such as toys, paper, furniture, and even whole buildings. Your home is probably made mostly of wood. You might eat at a wooden table or read this book in a wooden rocking chair.

But your house, toys, tables, and chairs weren't always there. The wood they are made of came from living trees! Those trees started as tiny seeds and grew for many, many years before they were large enough to *harvest*.

Harvesting trees, or *timber*, is a tough job, and often dangerous. After harvesting, the wood is hauled to a mill to be made into boards, also called *lumber*. From there, most companies will transport and sell the lumber to be used for buildings, furniture, and many other useful items. But Urban Lumber Company in Eugene, Oregon, is unique, because they harvest their own timber and work with it from beginning to end. Let's follow one part of their operation as they work to turn a tree into a lovely set of wooden chairs.

Preparing the Logging Site

And so the work begins! The people who harvest trees are called loggers. Before a logging company can get to work, they first purchase the timber. Often, the road leading to the timber is in very bad condition, so it has to be repaired to make sure the loggers can get safely to and from the *logging site*. Sometimes, there is no road at all, so one has to be made. This is a big job all by itself. Once the roads are complete, it's time for the trees to be harvested!

bulldozers are used to make dirt roads

roads leading to the logging site

the face

Douglas Fir Tree

TIMBER!

The man who brings the trees down is called a *faller*, and he gets busy first. Using a large, heavy chain saw, he makes a cut that only goes partway through the tree. Then he makes a second cut called an undercut so that a wedge-shaped, triangular chunk falls out of the tree trunk. This is called the *face*. This is important because it helps the tree fall right where the faller wants it.

On the other side of the trunk, the faller makes a cut just a little higher than the face. This is called the back-cut. The tree then loses its balance and falls in the direction of the face. A crisp crackling sound is heard as branches brush and break on their way down, louder and louder until . . . *Boom!* The giant tree hits the ground.

limbing – cutting the branches off of the trees

buck sawyer

All of a sudden, it's quiet—but only for a moment. Then the faller revs his chain saw back up and quickly trims the branches off the tree. This process is called *limbing*. Next, the log is cut into lengths that can be hauled to the mill. Loggers call this bucking, and the logger who does the bucking is called the *buck sawyer*.

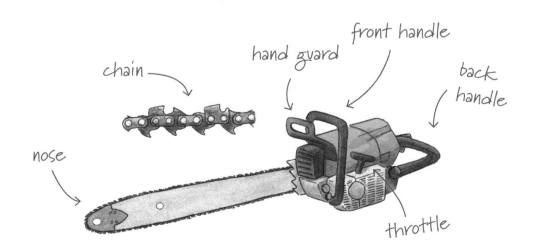

← the landing

The logs must then be moved to the *landing*, which is a flat area near the road. Here, they are piled up, ready to be loaded onto a truck. A machine called a *cat* or *skidder* can go right into the brush and move logs that are easy to reach, but if the logs are in an area that is steep or hard to get to, a yarder is used to pull them to the landing.

moving logs to the landing

SETTING UP THE YARDER

A *yarder* is a huge tower set up near the landing, with three to eight strong cables connected to it. These cables are called *guy wires*. They run from the top of the tower to the ground and keep the tower from tipping over. Another strong cable runs from the top of the tower to wherever the hard-to-reach logs are. This is a main line, also known as the *sky line*. A motorized *carriage* runs on the sky line, and another cable called the *skidding line* goes through the carriage. Logs are hooked to the skidding line, and then the carriage can pull the skidding line with the hooked logs all the way to the landing.

logger climbing a tree to hang a block for the skyline

In order for this to work well, the sky line on the yarder needs to be very high up. How do they keep this main line high up in the sky? Well, first, one of the loggers climbs a tree—sometimes up to a hundred feet. That's about the height of a ten-story building! Besides climbing dangerously high, he might also carry a long rope, an axe, and a chain saw too! Once he is high enough, he saws off the top part of the tree so that he can hang a block. Then the sky line is threaded through the block. The block holds the sky line up, so that the carriage and skidding line have enough room to work. Cables also have to be connected to this tree and anchored to hold it back securely during the operation.

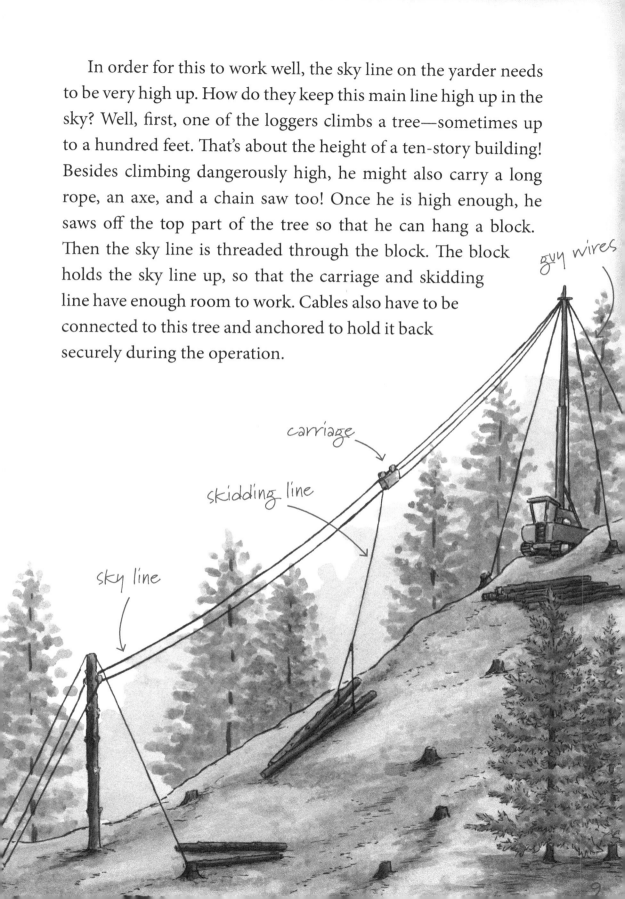

Moving the Timber

With the yarder set up, the *choker-setters* get busy. The choker-setters' job is to set a special looped cable end called a *choker* around the thick end of the log. They then connect the hooked logs to the chokers on the skidding line.

The head choker-setter is called a *rigging slinger*. He tells the loggers which logs to hook up to send to the landing. He also watches out for the safety of the workers.

Loggers communicate using radios called *talkie-tooters*, which whistle or beep. They use these to talk to each other in code. The man running the yarder needs to know when the logs are set and ready to move, so the rigging slinger uses the code on the talkie-tooter. *Beep-beep* means "Go ahead on the skidding line." Four beeps say to loosen the cable. One beep to the yarder, and everything stops.

Talkie-Tooter Code
beep x2 = go ahead
beep x4 = loosen cable
beep x1 = STOP

choker-setters

chaser

log loader loading logs onto truck

When the rigging slinger sees that the choker-setters are done with the logs and in a safe place, he signals, and the yarder gets to work. With the tower up and the lines held high, the hooked logs can turn and move without damaging living trees.

Once the logs are on the landing, a man called a *chaser* unhooks the chokers. Now the *log loader* does its job. Grabbing the logs with a big claw like a monster, this huge machine picks up the logs and places them neatly onto the log truck.

THE SAWMILL

The log truck driver will now haul the logs from the landing to a sawmill. The logs on his truck are usually thirty-two to forty feet long. Often, he travels on steep, narrow, dangerous roads. Through the forest and into town, the log truck driver rumbles along with his heavy load until finally, he arrives at the sawmill. The truck is unloaded, and the logs are sorted right away. Another big job begins!

log decks

Sawmills are noisy and bustling with the business of changing the logs into boards. The mill workers operate big machines and keep them running smoothly. This is hard work, involving a lot of walking, climbing, and lifting.

barker infeed deck

barker

The first machine used is called a *barker*, because it takes the bark off the logs. Next, the logs go over to the *bucksaw*. The bucksaw has an enormous blade that cuts the logs to the correct length for the main machine, which is called the *head rig*. The bucksaw also cleans off any dirt and gravel that might be clinging to the logs.

Next, it's time for the head rig to come into action! This giant saw cuts the log again, this time into big *slabs*. This is when the boards begin to take shape.

head rig cutting slabs

At this point, not all of the edges have been cut, and some may still have a rounded shape. When the board is at this stage, mill workers call it a *cant*. The cants are now ready for the next machine, the *edger*. Here, the rounded edges are cut to make them square. The edger also rips the cants into smaller boards.

edger

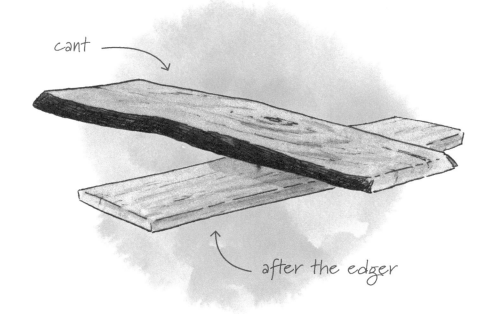

cant

after the edger

But the boards are not done yet! The lumber will be sawed again by a *resaw*. The resaw splits the board into two pieces, bringing the boards to their final size. A final machine called a *planer* takes the rough edges off the boards and makes them smooth.

planer

lumber grader

lumber drying

GRADING AND DRYING

Several times throughout this process, the logs and lumber are *graded*. This means the lumber is examined to determine its strength and appearance. This way, the boards can be organized according to what they are best suited for. For example, some boards are very strong, which makes them perfect for outdoor construction.

Boards with a lot of knots and defects are given a lower grade. Lower-grade boards are often used to make pallets, stakes, or fencing. They may also be used to help reinforce concrete.

The higher-quality boards are the ones with the nicest grain and the fewest marks. These boards are the most expensive and better suited for fine furniture.

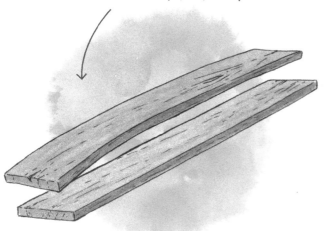

wood will warp if it does not dry properly

lumber ready for shipment

Once the wood is milled, the lumber needs to be stacked so that it can air-dry, a process that can take a few months. In summer, the wood dries much quicker. But if moisture leaves the wood unevenly, the wood will *warp*, meaning the boards may bend or change shape. To avoid this, some boards are put in a huge oven, called a *kiln*, to dry them out all the way. The kiln also helps to preserve the wood and clear out any termites or other insects that might have been hiding in it.

From here, the wood can be prepared for shipment, loaded onto big trucks or railcars, and sent to factories and furniture makers all across the country. So now you see how logs go into a sawmill and leave as boards!

SAWING AND SANDING

At Urban Lumber Company, the trees have been harvested, milled, processed, and dried, and are finally ready to be made into furniture. Now let's follow how their lumber is used to craft a set of beautiful chairs.

The new boards have many scrapes and bumps, so first, the workmen must use a *sander* to smooth the wood down to the thickness needed. After sanding, the lumber goes to a machine called a *straight-line rip*. Does it sound like this machine rips the boards in a straight line? If you guessed that, you are right! This machine is used to rip, or cut, straight edges. It can cut boards to many different lengths and widths. Sometimes, pieces of wood are glued together at the edges to create wider boards. The straight-line rip saw is used to make the boards the correct size for each furniture piece.

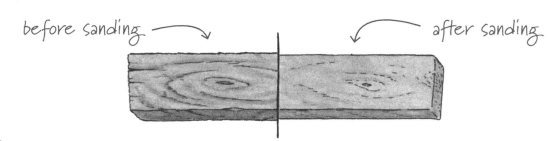

before sanding — after sanding

straight-line rip

Next, the boards are sanded again, now with a wide belt sander. The sanding on these chairs begins with 36-grit sandpaper. The particles in this sandpaper are large, which makes the sandpaper very rough. It smooths the scratchy surface of the wood down. Next, an 80-grit sandpaper, which is less coarse, is used. The last grade of sandpaper is a fine 120-grit, and this sanding evens out the boards and makes them smooth to the touch.

big rolls of sandpaper

wide belt sander

rough chair parts

Now free of lumps and bumps, the sleek wood can go to the next machine: the *router*! The router is an amazing tool that can be used to make many different furniture parts very accurately. First, small cutting tools called *bits* are attached to a round disc. The router turns the bits very fast to cut a *groove* or a design in an area of the wood. The bits come in many different shapes and sizes, so they can be used to make all sorts of patterns and pieces.

After being cut by the router, the pieces are sanded by hand. Any marks made by the machine are now smoothed away, and the surface of the wood is perfectly polished.

Notice the holes in the wood here? This type of cut is called a *mortise*. These mortises were also cut by the router. They are for structural joints. The workman makes sure that all the pieces are cut so that they will fit together perfectly, like a big puzzle. These jointed parts will make the final product much stronger.

mortises cut

ASSEMBLY

But after all that cutting, sanding, routing, and maybe even some gluing, the pieces that were made are still just pieces—parts of a whole. Now the workman can put them together.

First comes *sub-assembly*. During this process, these individual pieces of wood are fitted together to form the basic parts of the chair.

After sub-assembly, a *varnish* is applied with a spray gun. This treatment gives the wood a hard, protective coating. Sometimes when a darker color is preferred, a special solution called a *stain* is applied on the wood's surface. Stain is a liquid with a dye in it, which is used to color wood or bring out a grain pattern. However, these chairs are sprayed several times with a clear coat, so the natural beauty of the wood can be easily seen.

glued chair sides

worker spraying varnish

varnished chair backs

parts ready to assemble

The final *assembly* takes place with one more round of gluing, and all the pre-finished parts are put together.

Depending on the type of chair being made, there may be more steps to get this big job done. Some chairs may have more detail, such as designs carved in the wood. Others may require upholstering or may be painted. A rocking chair would need more parts added to the legs. But these chairs are complete at last and can now be delivered to a store for sale.

finished chairs

HAVE A SEAT

So the next time you sit in a wooden chair, at the dinner table or on a porch, to read a book or just to rest your feet for a bit, take a moment to think of all the hard work that brought that chair to you. Remember the rugged loggers and the hard-working mill workers. Think about the brave truck drivers and the busy furniture factories. But most of all, remember the forest. Your wooden chair has come a long way—it was once a living tree!

GLOSSARY of TERMS

assembly: To put parts together.
barker: A machine used for removing bark.
bit: A small cutting tool which is attached to a drill or router used for woodworking. There are many different sizes and shapes of bits.
bucking: Cutting logs into lengths required by the mill.
bucksaw: Large saw used to cut timber.
buck sawyer: Someone who specializes in bucking logs.
cant: Log that has been cut, but usually has one or more rounded edges.
carriage: A two-wheeled unit that rides on the skyline of a yarder and carries logs.
cat: A huge gas-powered vehicle, also called a bulldozer or a logging tractor. Cats run on ribbed treads that grip the ground. Loggers use them for loading or hauling logs and call them "cats" for the Caterpillar Company that makes them.
chaser: Name for a logger who unhooks chokers on the landing.
choker: A cable that is fastened around a log so it can be moved to the landing.
choker-setter: The person who attaches the chokers around the logs. This is a very hard and dangerous job.
edger: Machine that saws cants into smaller boards; also squares the edges.
face cut: The notch cut into a tree on the side toward which it will fall.
faller: Person who cuts down trees.
grading: Work in which a person examines lumber or wood products. A grade is assigned according to an established set of rules.
groove: A narrow channel or cut made with a tool.
guy wires: Strong cables that keep a yarder from tipping over.
harvest: The work of cutting down and gathering mature trees to be processed into lumber.
head rig: Huge machine that takes big logs and cuts them into cants. It is usually the principle saw at a mill.
kiln: A large oven used for drying lumber.
landing: Place to which logs are moved before being loaded onto a truck.
limb: 1. To cut branches off of a log or tree. 2. A branch of a tree.
logging site: An area where timber is harvested.
log loader: A machine that uses a large claw to load logs onto the logging truck.

lumber: A wood product, from the logs of trees, made by sawing and planing.
mortise: Small cuts and notches made in furniture pieces that allow them to be fitted together like a puzzle.
planer: A machine used to smooth the surface of rough lumber.
resaw: A piece of equipment that is used to split one board into two.
rigging slinger: The head choker-setter.
router: A woodworking tool that can be used to cut and shape wood in many different ways, especially to make grooves or designs. It is usually considered the most versatile tool in a wood shop.
sanding: When sandpaper of different grits, or levels of coarseness, is used to smooth and polish rough wood. Can be done by hand or with machines.
skidder: A logging machine that can move or load logs. It usually runs on big heavy-duty tires.
skidding line: A cable that is attached to the carriage of a yarder.
sky line: Also called the main line; the cable that is strung between the yarder and the top of a tree at the logging site.
slab: A log that has been cut by a saw, but has not yet been cut into boards.
stain: Color added to wood, which changes it from its original natural color.
straight-line rip: A machine which cuts lumber in straight lines.
sub-assembly: The beginning of putting parts together.
talkie-tooter: A radio device (often carried on a belt), used by loggers to communicate in code with each other and with the yarder engineer.
timber: 1. Standing trees. 2. Lumber in a certain size class; may also be known as beams, girders, etc.
varnish: A protective coating that hardens when dried and protects wood from later damage.
warp: When wood dries unevenly and loses its shape.
yarder: Machine with cables that run from the top of a steel tower into the woods. Logs are hauled through the air on cables, to the landing. Also known as a tower skidder.

SPECIAL THANKS GOES TO:

Brad Hatley
Nick Johnson, cdlumber.com
Gerry Moshofsky, newood.com
Larry Rook
Seth SanFilippo, urbanlumbercompany.com

With deep gratitude for your unfailing help with my many questions. Your time and wisdom were essential to the writing of this story.

AND TO:

Mary Baird
Joseph Carnes
Nanci Carnes
Kendra Densmore
Rachel Ellis
Rex and Debbie Fuller
Joanne Gordon
Guy and Shelley Maynard
James McDonald
Cathy SanFilippo
Mike and Eileen SanFilippo

Philip SanFilippo
Steve and Teresa SanFilippo
Rick and Sandy Smith
Michael and Ruth Stafford
David Terry
Dolores Way
Paul and Roseanne Webb
James and Viva Wheaton
Cassidy Winters
Elona Wong
Maria Yager

Thank you all sincerely for your generous financial support.
Your kindness is greatly appreciated.

Photos Contributed by:

Bart Blair
C and D Lumber
Daniel Crume
Jason Faught
Thomas Gallagher
Daniel Gamet
Eric Krume

Juniper McDonald
Roseanne McDonald
Zack Pickens
Austin Sedy
Kyle Tomasovich
Urban Lumber Company
Viva Wheaton

Rose and her grandchildren

THE AUTHOR

Roseanne McDonald lives in the rural countryside of Oregon. Although she always wanted to be a writer, she set that dream aside to raise five children and pursue a career teaching elementary school. Finally, at the age of sixty-five, she is ready to share her stories with the world!

THE ILLUSTRATOR

Mary Barrows is a freelance illustrator from the small town of Walkersville, MD. Since she was old enough to hold a crayon she has been drawing pictures of her favorite stories, and she hasn't stopped yet. Her illustrations are done in a traditional style using ink, watercolor, gouache, colored pencil, graphite or whatever else she can get her hands on. To contact, visit: www.marybarrowsillustration.com

CPSIA information can be obtained
at www.ICGtesting.com
Printed in the USA
LVHW071455181219
640936LV00023B/460/P